The Unified Theory of Mass, Gravity, Light & The Big Bang

By

Joseph L. Poggie and Robert A. Poggie, Ph.D

ISBN: 1-4033-7086-9 (e-book)
ISBN: 1-4033-7087-7 (Paperback)

Library of Congress Control Number: 2002094239

This book is printed on acid free paper.

Printed in the United States of America
Bloomington, IN

1stBooks - rev. 03/28/03

Table of Contents

INTRODUCTION

Everyone is familiar with the force of attraction that the earth exerts on all objects near it and causes them to fall toward the earth's center. If the object is restrained from falling by a spring, which is calibrated in force versus deflection, a scale is created that measures the force of the earth's gravity upon the object, otherwise known as the weight of the object.

Isaac Newton was the first person to recognize and to organize the effects of gravity. He formulated the law of Universal Gravitation that states:

> *Every particle of matter attracts every other particle with a force*

that is directly proportional to the product of their masses and is inversely proportional to the square of the distance between them.

A cursory examination of Newton's Law of Universal Gravitation, however, leaves the reader with a sense of disappointment. One would expect something more substantial and physically tangible than an empirically derived statement, especially from the person who first accurately defined the laws of motion. Examination of Newton's Law of Universal Gravitation suggests several questionable areas. For example:

1) The law does not suggest any definable physical law or theory

2) It cannot be derived from any law or theory.

3) The dimensional analysis of Newton's Gravitational Constant (G) is compatible with circular, orbital motion, (as perceived by Kepler), but not with linear motion (as in the Cavendish experiments). Nevertheless, Newton combined the two into one theoretically illogical conclusion (as will be proven later).

4) Newton's Law of Gravity may be restated in a form that is mathematically identical to Newton's original Law. The restated version, however, has the more logical appearance of being a fragment of an equation for the addition of two forces.

5) Based on the restated version of Newton's Law, (and by extension to Newton's original law), gravity does not appear to be an attracting force.

instead of acting to slow the expansion of the universe, <u>gravity actually appears</u> <u>to be generated by the expansion of the universe</u>. This implies that antigravity does not exist and therefore cannot be the force that is causing the universe to expand.

In the following chapters, we will review and restate Newton's Law of Gravitation, and propose an alternative, new theory of gravity. And, through this reasoning, a new theory of mass is offered, which is based on the substance of dark matter. The theory of light transmission and Maxwell's equations are also reviewed, and a theory of light transmission based solely on waves propagating within and through dark matter is proposed. And lastly, this new theory of gravity is compatible with current theory that the universe is expanding as a natural result of the big bang, and that its expansion does not appear to be slowed by gravity.

CHAPTER I Newton's Law of Universal Gravitation

Newton based his Law of Universal Gravitation on Tycho-Brahe's observations of the position of the planets as they revolve around the sun and on Kepler's analysis of these same observations. Kepler found that the velocity of all planets, as they revolve around the sun, is inversely proportional to the square root of the distance between the sun and the planet, that is;

$$v \approx (1/R)^{1/2}$$

By the application of Newton's Second Law of Motion, the sun attracts each planet with a force (**F**) that is proportional to the mass (**m**) of the planet

$$F \approx m.$$

Combining his second Law of motion with Kepler's analysis, Newton concluded that a planet moving in a circular orbit around the sun, at a velocity (**v**), has an acceleration (**a**) equal to v^2/R toward the sun. It follows, therefore, that the sun must attract each planet with a force such that:

$$F = ma = m\ v^2/R.$$

Newton felt that it was reasonable to expect that this force relationship between the sun and the planets could be extended to include the force between the planets. Building on this hypothesis, Newton extended Kepler's analysis to include all bodies and thus formulated his Law of Universal Gravitation.

Expressed mathematically, Newton's Law of Universal Gravitation states that the attractive force (F_G) between any two bodies is:

$$F_G = G\, \frac{m_1\, m_2}{d^2} \qquad [\,I - 1\,]$$

where m_1 and m_2 are the masses of the bodies, d is the distance between them, and G is a gravitational constant.

In engineering units:

$$F_G = \text{lbs.}$$

$$m = \frac{lbs - sec^2}{ft}$$

$$d = ft$$

$$G = 3.44 \times 10^{-8}\ ft^4\ lb^{-1}\ sec^{-4}$$

The value of G was determined in a laboratory by Cavendish with the apparatus represented in FIGURE I - 1. Two small, light, spheres of equal mass (m_1) were attached to the ends of a light rod.

A quartz filament suspended this assembly, which had a small mirror attached as shown. Two heavy spheres, each of mass m_2, were mounted on a frame so that they could be moved toward the small spheres. When the two large spheres were moved toward the two small spheres, to the relative positions shown, the quartz filament was twisted in one direction. When the spheres were shifted to the diametrically opposite position, (shown in dashed lines) the filament was twisted in the opposite direction. The angle through which the filament was twisted was measured by observing the deflection of a light beam reflected off the mirror. With the dimensions of the apparatus and the physical properties of the filament known, Cavendish was able to determine the value of G.

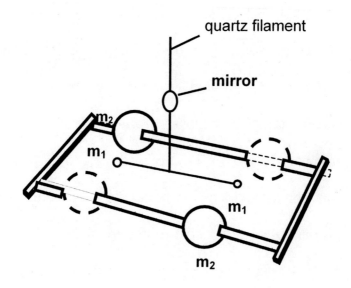

FIGURE I -1

CAVENDISH EXPERIMENT

CHAPTER II Newton's Law of Universal Gravitation Restated

It is obvious that Newton's Equation for Universal Gravitation (equation [I-1]) does not exhibit any sense of logic other than the possibility that it is the product of curve matching. Since curve matching, alone, rarely yields actual complete physical law or possible theoretical concept equations, it becomes necessary to examine the relationship of the original empirically derived equation as compared to proven laws of physics.

In order to make such a comparison, we begin by considering Newton's original equation [I -1] and inserting the dimensional units of G (in place of G). Newton's Equation may now be validly expressed as:

$$F_G = 3.44 \times 10^{-8} \; ft^4 \; lb^{-1} \; sec^{-4} \; \frac{m_1 m_2}{d^2} \qquad [II-1]$$

or as,

$$F_G = 3.44 \times 10^{-8} \; lb^{-1} \; \frac{ft^2}{sec^2} \; \frac{ft^2}{sec^2} \; \frac{m_1 \, m_2}{d^2} \qquad [II-2]$$

or as, $F_G = \dfrac{3.44 \times 10^{-8}}{lb} \; \dfrac{m_1 v^2 \, m_2 v^2}{d^2}$. $\qquad [II-3]$

In terms of force instead of in terms of mass:

$$F_G = \frac{3.44 \times 10^{-8}}{F_X} \; F_1 \, F_2 \; . \qquad [II-4]$$

Apparently, the **lb^{-1}** dimension (**F_X**) was included in the dimensional units of **G** in order to balance the dimensional units on both sides of the equation but the identity of **F_X** is not evident. In the special case, however, where the two bodies possess equal mass (**$m_1 = m_2$**), **$F_2 = F_1$** and **F_X** must also be equal to **F_1** Logically, equation [II - 4] may now be restated as

7

$$F_G = 3.44 \times 10^{-8}\ F_1$$

or as $$F_G{}^2 = [11.83 \times 10^{-16}\ F_1{}^2\ ]$$

and, when F_1 and F_2 are not equal, we may logically assume

$$F_G = 3.44 \times 10^{-8}\ [F_1\ F_2]^{\frac{1}{2}}. \qquad [II - 5]$$

Thus, in terms of mass, Newton's original equation [I - 1] may be restated as

$$F_G = G\ \left[\frac{m_1\ m_2}{d^2} \right]^{1/2} \qquad [II - 6]$$

which resembles a fragment of an equation for the vector sum of two forces[1]

---

[1] The vector sum (F_G) of two forces (F_1 and F_2)

$$F_G = (F_1{}^2 + F_2{}^2 \pm 2F_1F_2)^{1/2}$$

in terms of mass, instead of force:

$$F_G = \left[\left(\frac{G\ m_1}{r}\right)^2 + \left(\frac{G\ m_2}{r}\right)^2 \pm \frac{2G^2\ m_1\ m_2}{r^2} \right]^{1/2}$$

8

Due to the fact that the combined dimensional units of Newton's original Gravitational constant are meaningless, the exact nature of Newton's constant is not evident. It may be possible, however, to obtain a clue to help us understand it by means of the simultaneous solution of Newton's equation and the standard equation for the vector sum of two forces,

Starting with Newton's original empirical equation:

---

or, when $m_1 = m_2$, $F_G = G \dfrac{m_1 \text{ (or } m_2)}{d}$

where G is equal to 3.44×10^{-8} ft^4 sec^{-4} and

$$F_G = \frac{m_1 \, v^2}{d}$$

where $v^2/d = v^2/r = a = 3.44 \times 10^{-8}$ ft sec^{-2}.

$$F_G = G \frac{m_1 m_2}{d^2}$$

$$F_G = \frac{3.44 \times 10^{-8} \text{ ft}^4}{\text{lb} - \text{sec}^4} \frac{m_1 m_2}{d^2}$$

$$F_G = \frac{3.44 \times 10^{-8} \text{ v}^4}{\text{lb}} \frac{m_1 m_2}{d^2}$$

and expressing it in terms of force, instead of in terms of mass

$$F_G = 3.44 \times 10^{-8} \frac{F_1 F_2}{F_X} \text{ (say } F_2)$$

$$F_G = 3.44 \times 10^{-8} F_1 \qquad [\text{II - 8}]$$

Solving the resulting equation [II - 8], simultaneously with the standard expression for the vector sum of two forces (equation [II - 9])

$$F_G = [F_1^2 + F_2^2 + 2F_1 F_2 \cos \theta]^{1/2}, \qquad [\text{II - 9}]$$

we obtain:

$F_G = 3.44 \times 10^{-8} \; F_1 \; = \; [\; F_1^2 \; + \; F_2^2 \; + 2F_1 F_2 \cos \theta \;]^{1/2}$

$F_G = \; 3.44 \times 10^{-8} \; F_1 \; = (F_1 \; + \; F_2) \quad \text{when} \; \theta \; = \; 0^0$

$F_G = 3.44 \times 10^{-8} \; F_1 \; = \; (F_1 \; - \; F_2) \; \text{when} \; \theta \; = 180^{\,0}$

$$\text{Let} \; F_2 \; = \; kF_1$$

$3.44 \times 10^{-8} \; F_1 = [\; F_1^2 \; + \; K^2 \; F_1^2 \; \pm \; 2kF_1^2 \cos \theta \;]^{1/2} \; [\text{II - 10}]$

$3.44 \times 10^{-8} \; F_1 \; = \; F_1 \; \pm \; kF_1$

$3.44 \times 10^{-8} \; F_1 \; = \; F_1 \, (1 \; \pm \; k)$

$3.44 \times 10^{-8} \; = \; 1 \; \pm \; k$

When $\theta = \; 0°$, $k \; = 3.44 \times 10^{-8} - 1.0 = -1.0$ (essentially)

When $\theta = 180°$, $k = 1 - 3.44 \times 10^{-8} = +1.0$ (essentially)

If we first substitute +1 for k, equation [II - 10] becomes:

$3.44 \times 10^{-8} \; F_1 \quad = \; (F_1^2 \; + \; F_1^2 \; + 2F_1^2)^{\;1/2}$

$3.44 \times 10^{-8} \; F_1 \quad = \; F_1^2 \; + \; F_2^2 \; + \; 2F_1F_2)^{1/2}$

$3.44 \times 10^{-8} \; F_1 \quad = \; F_1 \; + \; F_2$

Since $k = F_2/F_1 = +1$,it is clear that $F_2 = F_1$ and $F_G = 2F_1$

Thus, for the case where $k = +1$, the above can only be true if Hypothesis A (below) applies:

Hypothesis A: The two forces are <u>internal attracting forces that tend to pull m_1 and m_2</u> together, as described in the following chapter in Figure III - 1 (vector sum - attracting forces - non-interacting) and Figure III - 2 (vector sum - attracting forces - interacting).

In a similar manner for the development of Hypothesis A where $k = +1$, by setting $k = -1$, equation [II - 10] becomes:

$$3.44 \times 10^{-8} \ F_1 \quad = \quad (F_1{}^2 + F_1{}^2 - 2F_1{}^2)^{1/2}$$

$$3.44 \times 10^{-8} \ F_1 \quad = \quad (F_1{}^2 + F_2{}^2 - 2F_1F_2)^{1/2}$$

$$3.44 \times 10^{-8} \ F_1 \quad = \quad F_1 - F_2$$

Since $k = F_2/F_1 = -1$, it is clear that $F_2 = -F_1$ and is in accordance with Newton's Second Law of motion. Thus, for the case where $k = -1$, the above can only be true if one of the following Hypotheses is true (Hypotheses B, C or D):

Hypothesis B: If F_1 and F_2 are <u>internal forces pushing the two masses apart</u>. This solution is obviously irrelevant since gravity forces masses together, not apart.

Hypothesis C: If the two forces are <u>internal forces that react on each other to pull m_1 and m_2</u> together, as described in Figure [III - 2].

Hypothesis D: If the two forces, (F_1 and F_2), are <u>external forces pushing m_1 and m_2 together</u> as described in Figure [III - 3].

SUMMARY

1) **Newton's equation for his Law of Universal Gravitation, in its original form, states that**

$$F = G\ \frac{m_1\ m_2}{d^2},$$

where

$$G = 3.44 \times 10^{-8}\ ft^4\ lb^{-1}\ sec^{-4}.$$

2) Newton's equation is easily re-stated to read

$$F = G\left[\ \frac{m_1\ m_2}{d^2}\ \right]^{1/2}$$

where $G = 3.44 \times 10^{-8}\ ft^4\ sec^{-4}$.

3) In its original form, Newton's equation appears to be an empirically derived statement with no clear basis in definitive physical law or theory. However,

14

in its re-stated form (equation [*II-6*]) it <u>resembles a fragment of an equation for the vector sum of two forces</u>.

CHAPTER III Newton's Law Of Universal Gravitation Expanded

In the previous chapter, Newton's Gravitational Law was reformulated and four hypotheses put forth on the relationship between the vector forces and interacting masses. With the restatement of Newton's Law, and despite the fact that it is universally accepted as the formula that quantifies the physical forces acting between two or more bodies in space, it is clear that it does not suggest a definable physical law or theory that explains this action. In this chapter, we will examine each of the hypotheses and suggest an alternative theory and origin of gravity.

The most obvious reason that Newton's Law must consist of two interacting bodies is the fact that logically, regardless of the actual nature of gravity, the gravitational force (F_G) between any two bodies must consist of two forces. Newton, himself, reasoned that each body exerts a force on the other; and not where only one body exerts a force on the other body. Since each force is a vector quantity, having the dimension of direction as well as of magnitude, it follows that the resultant total gravitational force must be the vector sum of the two forces (F_1 and F_2). Each of the forces is assumed to be generated by one mass, and acting on the other mass. Since the forces would be exerted along the same line of action, and since they would both tend to force the two masses together, it would appear that they would be additive, that is, $F_G = F_1 + F_2$.

Newton's original equation [I - 1,] and its more logically restated version, equation [II - 6], appears to somewhat resemble the fragments of the cosine law equation for the vector sum of two forces (F_1 and F_2). The accepted explanation of Newton's Law of Universal Gravitation is that both forces are defined as attracting forces acting within a system consisting of the two masses m_1 and m_2. The reason for this explanation is derived in Hypothesis (A) of the simultaneous solution in Chapter II, and as illustrated below in FIGURE III − 1.

FIGURE III - 1   *VECTOR SUM -*
ATTRACTING FORCES
NON INTERACTING

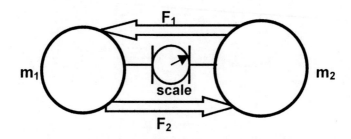

FIGURE III – 1A

FIGURE   III - 1A: Mass m_1 exerts a **traction** force (F_1) on mass m_2 while mass m_2 exerts a **traction** force (F_2) on mass m_1 . The total **traction** force, exerted on the connecting scale, is known as the gravitational force (F_G)

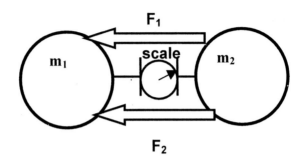

FIGURE III – 1B
Simplified equivalent of FIGURE III - 1A

FIGURE III - 1B: both forces tend to pull **m₁** and **m₂** together. The total gravitational traction force (F_G) between the two masses is simply:

$$F_G \;=\; F_1 \;+\; F_2 \,. \qquad\qquad [\text{III-1}]$$

By the process of vector addition, equation [III - 1] becomes

$$F_G = [\, F_1^2 + F_2^2 + 2 F_1 F_2 \cos \theta \,]^{1/2}.$$

Since the angle (θ) between F_1 and F_2 is **0°**,

cos θ **= 1**:

$$F_G^2 = [F_1^2 + F_2^2 + 2F_1F_2]. \qquad [\text{III-2}]$$

Factoring equation [III -2] results in:

$$F_G = [F_1 + F_2]. \qquad [\text{III-3}]$$

Expressed in terms of mass, equation [III - 2]

becomes:

$$F_G = \left[\left(\frac{G\,m_1}{r}\right)^2 + \left(\frac{G\,m_2}{r}\right)^2 + \frac{2\,G\,m_1\,m_2}{r^2}\right]^{1/2} \quad [\text{III-4}]$$

and equation [III- 4] is easily factored to yield:

$$F_G = G\left[\frac{m_1 + m_2}{r}\right] \qquad [\text{III-5}]$$

where $G = ft^2\ sec^{-2}$, or v^2.

Newton's equation [I - 1], by virtue of its restated version, Equation [II -6], now appears to resemble the third quantity in the brackets of the equation for the vector sum of two forces, equation [III - 4]. The attracting force of gravity, however, as determined by the cosine equation for the vector sum of two forces (The square root of equation [III - 4]), still does not agree with the findings of Newton, of Kepler or of Cavendish since, in this case, it states that the total attractive force between two bodies is proportional to of the sum of their masses – not to the square root of the product of the masses. It also appears that there is no explanation of the constant (G) in Newton's equation since, as it stands, Newton's Equation deals with radial acceleration while Cavendish's experiment was conducted with masses that were stationary with respect to each other.

In view of the above question, it becomes necessary to examine Newton's empirical equation more closely in order to resolve the difference between theory and fact. Starting with the logical definition of all forces being vector quantities, we may proceed to convert Newton's original Law of Universal Gravitation:

$$F_G = 3.44 \times 10^{-8} \ ft^4 \ lb^{-1} \ sec^{-4} \ \frac{m_1 \ m_2}{d^2}$$

into it's nearest apparent equivalent in vector forces

$$F_G = 3.44 \times 10^{-8} \ F_1 \ F_2$$

and then expand into the standard form for the addition of vector forces, (which it appears to resemble),

$$F_G = [F_1^2 + F_2^2 + 2 \ [1.72 \times 10^{-8}] \ F_1 \ F_2]^{1/2},$$

we have an equation that cannot be factored. However, if we set F_1 equal to some multiple (k) of F_2, (F_1 = k F_2),

$$F_G = [k^2 F_2^2 + F_2^2 + 2[1.72 \times 10^{-8}] k F_2^2]^{1/2}.$$

Dividing by F_2^2 we have, for the right side of the equation,

$$[1 + k^2 + 2[1.72] \times 10^{-8} k]^{1/2}$$

Placed in the standard form for a second degree equation and set equal to zero,

$$[k^2 + 3.44 \times 10^{-8} k + 1.0] = 0.$$

Applying the quadratic formula:

$$k = \frac{-b \pm (b^2 - 4ac)^{1/2}}{2a} = -\frac{3.44 \times 10^{-8} \pm (11.83 \times 10^{-16} - 4)^{1/2}}{2}$$

where:

a = 1.0

b = 3.44×10^{-8}

$$c = 1.0$$

and evaluating the discriminant:

$$b^2 - 4ac = 11.83 \times 10^{-16} - 4$$

which turns out to be negative, and as such, reveals that the roots are imaginary and that there are no real solutions to this equation, regardless of the value of k. *This result indicates that the concept of gravity as an internal attracting force (Chapter II, Hypothesis A) may have to be revaluated.*

The second possibility to be ascertained from Chapter II, Hypothesis C, is that gravity may be the combination of two interacting, internal forces (F_1 and F_2), produced by the two masses (m_1 and m_2) in the system. In this case, the two forces are assumed to be acting, by some unknown means, on each other as shown in FIGURE III-2, not each

force acting on the opposite mass.

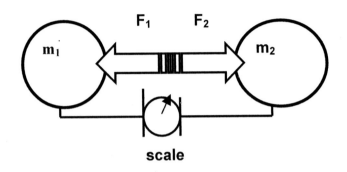

scale

FIGURE III - 2 VECTOR SUM - ATTRACTING

FORCES – INTERACTING

Force F_1, generated by mass m_1,

interacts with force F_2, generated

by mass m_2.. F_1 tends to pull m_2

towards m_1 while F_2 tends to pull

m_1 towards m_2

By virtue of Newton's Third Law Of Motion, (every action has an equal and opposite reaction),

$F_2 = -F_1$ <u>always.</u>

Since F_1 and F_2 are internal traction forces, both forces tend to pull mass m_1 and mass m_2 together.

Referring again to Newton's Third Law of Motion, F_1 and F_2 constitute a single action/reaction couple with F_1 (or F_2) being the action and F_2 (or F_1) being the reaction. F_1 MUST BE EQUAL TO F_2 – ALWAYS !!!. The attracting (gravitational) force between the two masses is simply F_1 (or F_2), in terms of mass:

$$F_G = G \ \frac{m_1 \ (or \ m_2)}{d} \quad \text{where } G = ft^2 \ sec^{-2}.$$

This hypothesis (C) has the appearance of some validity since it takes the form of $F_G = mv^2/r$. The problem is that this is the equation for curved motion, and we are dealing with linear motion. In addition, this solution is unacceptable since it does not account for the gravitational effect of the second mass. For example, water would have the same weight per cubic foot on the sun as well as on the earth — or on any other body in the universe.

The third and only remaining hypothesis from Chapter II involves gravity as being composed of external opposing forces as derived in Chapter II, Hypothesis D, and illustrated in FIGURE III - 3.

FIGURE III- 3

VECTOR ADDITION OF TWO

OPPOSING EXTERNAL FORCES

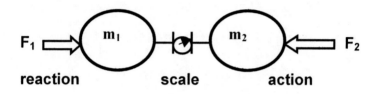

reaction   scale   action

FIGURE   III - 3A

(FIGURE III-3A). An external force (F_2), is exerted on mass m_2, which is resisted by a second external force (- F_1) exerted on mass m_1 .

The sum of the two forces on the m_1, m_2 mass system is the displacement force (F_D), which tends to displace the entire m_1, m_2 mass system.

$$F_D = F_1 + (-F_2)$$

Joseph L. Poggie & Robert A. Poggie, Ph.D.

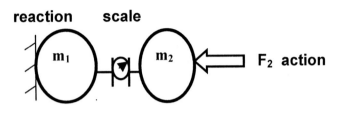

reaction   scale

m_1 m_2 F_2 **action**

FIGURE   III - 3B

(FIGURE III- 3B). The total force (F_G) pushing the two masses together, **is the smaller of the two forces (F_2).**

$$F_G = F_2$$

Starting with the displacement force

$$F_D = F_1 + (- F_2)$$

and expanding it into the standard form for the addition of two vectors:

$$F_D = [F_1^2 + F_2^2 - 2F_1(-F_2)\cos\theta]^{1/2}$$

$$F_D = [F_1^2 + F_2^2 + 2F_1 F_2 \cos\theta]^{1/2}$$

30

Since the angle $\theta = 180°$, $\cos \theta = -1$:

$$F_D = [F_1^2 + F_2^2 - 2F_1 F_2]^{1/2} . \quad [\text{ III - 6 }]$$

Factoring equation [III - 6] :

$$F_D = (F_1 - F_2)$$

$$F_G = F_2$$

where F_D is the sum of the displacement forces applied to the entire system (m_1 , m_2) and F_G is the (gravitational) force pushing the two masses together.

Expressed in terms of mass, equation [III- 6] becomes

$$F_D = [(m_1 a_1)^2 + (m_2 a_2)^2 - 2 m_1 a_1 m_2 a_2]^{1/2} . [\text{III - 7 }]$$

Factoring equation [III-7] and solving as in CHAPTER IV (equation [IV–5]), where we are working with mass density of the fluid through which the body is moving:

$$F_1 - F_2 = m_1 \, dv/dt - m_2 \, dv/dt$$

$$F_1 - F_2 = \rho A_1 \, dl/dt \, dv/dt - \rho A_2 \, dl/dt \, dv/dt$$

$$F_1 - F_2 = m_1 a - m_2 a$$

$$F_G = F_2 = g m_2$$

Since we are working with the mass density of the fluid (ether), it follows that the mass involved is equal to the mass density of the fluid x the volume of fluid displaced per unit of time, not the mass of the body

thus:
$$m = \rho A \, dL$$

and has dimensional units of

$$\frac{lb - sec^2}{ft^4} \times ft^2 \times ft \quad or \quad \frac{lb - sec^2}{ft}$$

where $g = a =$ the acceleration of the smaller mass due to gravity of the reference (larger) mass, (32.2 ft/sec^2 on the surface of Earth). Actually, both masses contribute to the gravitational force g and would be a function of both.

According to the hypothetical model that we have developed in chapter IV, MASS, the mass density (ρ) of the ether (as it travels toward the 'depression' caused by the masses) must be inversely proportion to the square of the distance between the masses making the gravitational force to be inversely proportional to the square of the distance, even though it is not readily apparent from the equation.

The concept of gravity as being composed of <u>opposing external</u> forces appears to be a more logical explanation of the force of gravity than does the concept of gravity as being composed of <u>internal attracting</u> forces. A major difference between the two is that, in the external force concept, the effect of the second mass is accounted for and explained, in detail, in

SECTION - <u>V Gravity</u>.

SUMMARY

1)   Newton's Equation for the Law of Universal Gravitation appears to consist of a fragment of an equation for the vector sum of two forces <u>but is not a fragment of the vector sum of attracting forces between two masses</u>.

2)   The perception of gravity as an internal attracting force does not appear to be valid.

3)   Gravity appears to be composed of external forces pushing bodies together.

CHAPTER IV Mass

Prominent in any theory of gravity is the mysterious property known as mass. Mass may be defined as that property which causes a body to possess inertia and is common to all matter. The concept of mass is so important that it is difficult to imagine solving most scientific or engineering problems without it. In spite of this, the nature of mass is a total mystery that must be understood before we can ever hope to understand or to control the force of gravity.

Isaac Newton provided us with virtually everything we know about mass today in his second law of motion.

NEWTON'S SECOND LAW OF MOTION

An unbalanced force (**F**), acting on a body, causes the body to accelerate in the direction of the force. The acceleration **(a)** is directly proportional to the unbalanced force and inversely proportional to the mass of the body. Mathematically:

$$a = F / m \qquad [\, IV - 1 \,]$$

By rearranging equation [IV - 1] so that it reads **m = F/a**, we obtain an expression that defines the mass of a body as the ratio that is obtained when the force applied to the body is divided by the resulting acceleration of the body. In this way we can measure and use mass any where in the universe, but we do not have a clue as to what causes it. In order to try to understand the nature of mass and the mechanics of its manifestations, let us rearrange equation [IV - 1] again so that:

F = m a   [IV-2]

and look for another expression for force that is also a function of non-uniform linear motion. The only expression that comes close is the equation that describes the mechanics of a body moving through a fluid at constant velocity:

F = 1/2 k ρ A v^2   [IV-3]

where:

> **F** = force on the body
>
> **A** = the cross-sectional frontal area of a solid body moving through a fluid
>
> ρ = the mass density of the fluid through which the body is moving
>
> **k** = shape factor (assume 1.0)

Since equation [IV -3] describes uniform linear motion and we want non -uniform motion, we take the derivative of equation [IV-3] with respect to time:

dF/dt = ρ k A v dv/dt   [IV-4]

Replacing **v** with **dL/dt** and multiplying both sides by **dt** we obtain

$F_2 - F_1 = \rho$ **k A dL dv/dt**   [IV-5]

Since ρ x **A** x **dL** = **mass density x volume = mass (m),**

$F_2 - F_1$ = **k m dv/dt**.

Substituting acceleration **(a)** for **dv/dt** and setting $F_1 = 0$, we have

F = k m a   [IV-6]

Now, however, the mass (**m**) refers to the mass of the fluid being accelerated by the motion of the body through the fluid, not to the mass of the actual body, as was the case with equation [IV - 2].

By virtue of Newton's Third Law Of Motion, "For every action there is an equal and opposite reaction," we find that

$$F = m_b \times a_b = m_F \times a_F \qquad [\text{IV-7}]$$

where:

    m_b = mass of the body,

    a_b = acceleration of the body,

    m_F = mass of the fluid,

    a_F = acceleration of the fluid,

this allows us to use the mass and acceleration of the body, which we are able to measure, in place of the mass and acceleration of the fluid which, in this case, we are not able to measure.

It now appears that the very existence of the property of inertia along with Newton's Third Law of Motion (equation [IV-7]) is a very strong indication, if not actual proof, of the reality of that mysterious, all pervading fluid ,"ether," which was hypothesized and then discarded a century ago. It also appears that we are not referring to bodies in the conventional sense. To do so would imply that all bodies of the same size and shape would be expected to have the same mass, regardless of their composition. This would also imply that all bodies, other than spheres, would exhibit different masses when the force is applied in different directions. This is clearly not the case since observation tells us that the mass of all bodies is a function of volume and composition and is independent of the direction of applied force.

But we also know bodies are composed of matter which consists mostly of empty space occupied only by essentially solid particles which are incredibly small and are tremendously widely spaced in all directions. If we postulate the existence of an all pervading "ether" which occupies and flows through these relatively vast open spaces in matter, unhampered except by the these solid particles, and if the solid particles are so widely spaced that, even if one immediately follows the other, the "ether" streamlines have both the time and distance to reform. We can then postulate that, in any body, the sum of the solid particles constitute the mass of the body and sum of the cross sectional areas of all of the particles make up the effective cross sectional area (A) of the body in equations [IV - 3], [IV -4], and [IV-5]. We now have the conditions that satisfy equation [IV - 6] and a theoretical model for mass that is necessary to develop a theory for gravity. Until now, Newton's Universal Law of Gravity worked because it is empirical, but no one has ever been able to derive it since there has been no theory upon which to base a derivation, and also because there is no explanation of its most prominent component - mass.

It should also be noted that both Einstein's theory of General Relativity and Newton's Law of Universal Gravitation are similar in that they each lack definitions of mass and that both theories postulate the existence of a mysterious field of force that requires no energy to maintain it. Einstein's Theory of General Relativity is based, in part, on the theory of Special Relativity, which has been proven to be to be in error (ref. 4), therefor, Einstein's Theory of General Relativity is discarded without further consideration.

CHAPTER V A New Theory For Gravity

In the first three chapters, Newton's law of gravitation was reviewed, its flaws discussed, and the equations logically reorganized for our purposes in developing a new theory for gravity. This new theory rests on Newton's original work, the new theory of mass proposed in the previous chapter, and observations made about the universe and its expansion. According to Hubbell, the universe appears to be expanding in accordance with the same laws that govern the expansion of solid bodies when the temperature is changed, that is, volumetric expansion is expressed as follows. For an isotropic body, a body that has the same physical properties in all directions, volume expansion (∇V) is essentially:

$$\nabla V = 3\,\alpha\,V\,\nabla T \qquad\qquad \text{[V-1]}$$

where: ∇V =   the change in volume (change in volume / unit volume /degree change in temperature)

α =   the coefficient of linear expansion

V = the original volume

∇T = the change in temperature.

When the solid body expands, the volume of any voids in the body will also expand as equation [V −1] and all dimensions (**L**) will vary as coefficient of linear expansion such that:

$$\nabla L = \alpha\,L = \nabla T \qquad\qquad \text{[V-2]}$$

If the cavity contains a gaseous fluid, the pressure of the fluid will vary as the inverse of the change in volume of the cavity

$$\Delta P \ = \ P \times \Delta T / 3 \, \alpha \, V \qquad\qquad [V\text{-}3]$$

Further, if the cavity contains some amount of another solid matter having a different coefficient of expansion (say zero) than the solid body containing the cavity along with the gaseous fluid, and the volume of the solid matter is some fraction (**k**) of the original cavity volume (**V**), then the change in the volume of the cavity (less the volume of the solid matter in the cavity) is:

$$\Delta V \ = \ 3 \, \alpha \times V \times \Delta T \ - \ kV \qquad\qquad [V\text{-}4]$$

and the change in pressure in the cavity is

$$\Delta P \ = \ P / [V(3 \, \alpha - k)] \, . \qquad\qquad [V\text{-}5]$$

If we:

(1) Replace the isotropic body with the universe.

(2) Replace the cavities in the isotropic body with the cavities in the universe that are occupied the Sun, Earth etc.

(3) In place of the solid matter and gas in the cavities of the isotropic body, substitute the Sun, Earth etc.

We now have a conceptual model of the universe. All bodies (such as stars, planets, people, and apples) in the universe occupy "holes" in the universe which tend to expand at the same proportional rate as the universe but the pressure of the ether inside the cavities decreases. The difference in pressure between the universe and the cavity of the earth causes a flow of ether (dark matter) into the earth from all directions. Any other bodies (people, apples, moons) caught in the path of the ether flow are forced toward the earth with a force (**F**) such that:

$$F = m\,a$$

Since gravity is apparently caused by the expanding universe and cannot exist unless the universe is expanding, it follows that:

(1) The 'big bang' could not have been the result of an explosion caused by the heat of compression of all of the original matter of the universe being compressed by a force of gravity that did not exist.

(2) The energy that powered the 'big bang' must have had another origin.

(3) The only other possible sources of energy are nuclear or impact (collision of two bodies).

(4) Nuclear fission would have required that all, or most of the elements would already have existed at the time of the big bang.

(5) Nuclear fusion would have required only the presence of Hydrogen atoms and some source of great amount of energy.

(6) A great amount of energy could have been supplied by impact, which would have required the presence of at least two massive bodies colliding at tremendous speeds in order to generate the enormous amount heat required.

(7) It would appear that a combination of items 5) and 6) is the most reasonable scenario.

Heat is transmitted in longitudinal compression waves. These waves, traveling in the ether of the universe, would appear to be the cause of the pressure that caused the universe, or any matter, to expand.

Other Possibilities

Base on the Doppler Effect alone, there are several possibilities:

(A) If we accept, as fact, that Einstein and Maxwell were wrong and that light really consists of longitudinal compression waves traveling in a conducting medium, the entire universe may be expanding, That is, all of the solid matter and all of the conducting medium (ether or dark matter) are moving outward. In this case only part of the observed spectrum shift would be due to the motion of the bodies, the rest of the shift would be due to the motion of the dark matter. If the solid matter and the conducting medium are traveling at the same velocity, the rate of expansion of the universe is one-half that presently calculated.

(B) The universe may not be expanding. Only the conducting medium, (dark matter) may be in motion while the rest of the universe is stationary. The Doppler effect is usually calculated, (for longitudinal compression waves) as occurring when the conducting medium is motionless and the source and/or observer are in motion. But an effect identical to the Doppler effect also exists when the source and observer are at rest with respect to each other and only the conducting medium is in motion.

In this chapter, we have proposed a new theory of gravity based on a differential in pressure of the ether (eg dark matter) found in bodies such as atomic nuclei, the earth and the sun, and the ether found in free space. Gravity is the result of a flow of ether from high pressure to low pressure.

The potential presence of an all-pervading ether offers many intriguing possibilities for revisiting existing theories and laws that involve the transmission of energy through free space. For example, the accepted theory of the dual nature of light (particle + transverse wave) could much more logically explained as a longitudinal wave traveling through the ether. The same logic also applied to the transmission of radio waves and heat through free space. In the following chapters, we will explore these possibilities.

CHAPTER VI Pressure and Heat

What is the mechanism causes the universe, or any other matter to expand? The laws of physics tell us that the addition of heat to any substance causes individual molecules in that substance to move about wildly and that these molecules, by virtue of their increased motion, require additional space. This additional space occupied by the heated molecules causes the matter to expand. If the heated matter is restrained from expanding; the result is an increase in force in the direction of restraint. In the case of a gas confined in a closed container, the force exerted by the excited molecules on the walls of the container constitute a given force per unit area or its pressure.

In theory, the volume of an unrestrained gas will increase infinitely or until the individual molecules are too far apart to contact each other, depending on the temperature of the gas. But what is the physical cause of the motion of the gas particles? Theory tells us that heat, itself, is simply molecules in motion. Since heat will flow from a hot item to a cooler item, it follows that some physical force must impart that motion to the molecules unless we attribute some weird sensory and muscular structure to molecules that causes them to sense the heat (what ever that is) and then react to it.

In engineering terms, we know that heat is transmitted from one place to another by:

a) Conduction, where the two bodies (the heat source and the heat sink) are in intimate contact with each other and the heat flows directly from the source to the sink.

b) Convection, where the heat source and the heat sink are separated from each other by a third medium which acts to obtain the heat, by conduction, carry the heat from the source to the sink by simple physical motion, and transfer it to the sink by conduction

c) Radiation, where the source and the sink are not connected by a conducting or by a convection medium. In this case the two bodies may be separated by nothing more than empty space.

It might be reasonable to postulate that, in a) conduction and in b) convection, the heat is transmitted directly from one material to the other by physical contact of the excited molecules. In item c) radiation, however this is not possible since there can be no physical contact between the source and the sink, directly or through an intermediate convection medium. Instead, it is easily proven that radiant heat energy is transmitted by 'electromagnetic' waves (similar to light) through empty space provided that one accepts as reality that, even in 'empty ' space, any wave action requires the presence of a wave-conducting medium.

Careful consideration of the available facts leads to but one possible conclusion; all heat energy, including conduction, is transferred by means of 'electromagnetic' radiation. But present theories of 'electromagnetic' radiation are based on the concept of 'transverse waves' that cannot be found in any phenomenon other than low amplitude vibrations in tightly stretched strings. The concept of transverse waves is, however, seriously flawed as is demonstrated in Chapter VII, Maxwell's Electromagnetic Theory of Light.

If we substitute the concept of longitudinal compression waves, (traveling through the all-pervading ether of the universe), in place of Maxwell's transverse waves, we have a completely logical solution to the problem of all types of heat transfer and of the resulting pressure changes.

CHAPTER VII Maxwell's Electromagnetic Theory of Light

Maxwell's contribution to the modern concept of the nature of light was the result of his attempt to generalize the principles of electromagnetism by applying his own mathematical analogy of Faraday's "Lines of Force" to the then known relationships of electric and magnetic action. The result was twenty basic equations which, when expressed in differential form, may be reduced to a new set of four equations that became the basis for the classical electromagnetic theory of light by showing that electromagnetic waves have all of the properties of light.

In the beginning, Maxwell did not attempt to construct a hypothesis that might lead to a theory of electromagnetism. He used instead the analogy of a fluid system containing an imaginary, frictionless, incompressible fluid as a tool by which he was able to adapt Faraday's "lines of Force", and his own abstract dynamical concepts, to relate and combine the known electric and magnetic actions into his four equations of electromagnetism. Because of this, there is no rigorous derivation of Maxwell's equations and his conclusions are valid only insofar as they agree with reality. For this reason, and also because a detailed explanation of Maxwell's analogies is immaterial to this book, we will examine only the four relations that Maxwell explored and his conclusions; regarding each which may have influenced Einstein's development of Special Relativity.

Maxwell's first equation is based on Gauss's Relation which states that the electric flux (Φ) passing outward through any closed surface equals the total electric charge **(Q)** inside the closed surface divided by the permittivity ($\mathcal{E}_0$) of free space.

$$\Phi = Q/\mathcal{E}_0. \qquad \text{[VII-1]}$$

Maxwell applied his own mathematical treatment of Faraday's "Lines Of Force" concept to Gauss's relation in order to obtain his own first equation:

$$\mathcal{E}\,ds = Q/\mathcal{E}_0 \qquad \text{[VII-2]}$$

which also states that the total electric flux through

a closed surface equals Q/ε_0 except that now an

imaginary, incompressible, frictionless fluid is an

analogy for electric flux in a tube having a cross

sectional area **(ds)** of some portion of the total

spherical area **(s)** surrounding a charge in space.

This allowed Maxwell to examine the electric field

in terms of fluid flow in the three mutually

perpendicular axis of a rectangular reference

system in space.

For our purpose, we may simplify [VII-2] by

setting **ds** equal to1.0 so that [VII-2] becomes:

$$\varepsilon = Q/\varepsilon_0$$

For the static electric field described in Gauss's

relation there can be no current flow and it follows

that **Q** is constant and therefore the change in flux

must be zero thus

$$d\mathcal{E} = 0 \qquad [VII - 3]$$

and the change in flux along each of the three

mutually perpendicular axis in space is zero, that

is;

$$\frac{d\mathcal{E}_X}{dX} + \frac{d\mathcal{E}_Y}{dY} + \frac{d\mathcal{E}_Z}{dZ} = 0$$

But, Maxwell reasoned, for a plane wave

traveling in the **X** direction, $\mathcal{E}_X$, $\mathcal{E}_Y$, $\mathcal{E}_Z$, are

functions of **X** and **t** only, so that the partial

derivatives of the these quantities with respect to

Y and to **Z** vanish leaving only

$$\frac{d\mathcal{E}_X}{dX} = 0 \quad \text{and} \quad \frac{d\mathcal{E}_X}{dt} = 0 \qquad [VII - 4]$$

as Maxwell's first equation

Maxwell's second equation states that the corresponding magnetic flux is also always equal to zero because there is no magnetic substance corresponding to the electric charge and therefore, by the same logic as the first equation, Maxwell's second equation becomes, in differential form,

$$\frac{d\beta_X}{dX} + \frac{d\beta_Y}{dY} + \frac{d\beta_Z}{dZ} = 0$$

Again, leaving out the partial derivatives with respect to Y and Z, we have Maxwell's second equation

$$\frac{d\beta_X}{dX} = 0 \quad \text{and} \quad \frac{d\beta_X}{dt} = 0 \qquad [VII - 5]$$

Maxwell's third equation is based on the law of magnetic induction which states, in effect, that a changing magnetic field (I) will always generate an electric field $(\mathcal{E})$ such that: $\mathcal{E} \, dl = - \, dI/dt$

$$\mathcal{E} \, dl = - (d\beta/dt)ds$$

where **dl** refers to the length of the electric field "tube" in Maxwell's analogy.

Maxwell stated this law, in differential form, as:

$$\frac{d\mathcal{E}_Z}{dY} - \frac{d\mathcal{E}_Y}{dZ} = \frac{d\beta_X}{dt}$$

$$\frac{d\mathcal{E}_X}{dZ} - \frac{d\mathcal{E}_Z}{dX} = - \frac{d\beta_Y}{dt}$$

$$\frac{d\mathcal{E}_Y}{dX} - \frac{d\mathcal{E}_X}{dY} = - \frac{d\beta_Z}{dt}$$

Leaving out all of the terms containing **d/dY** and **d/dZ** results in Maxwell's third equation

$$\frac{d\mathcal{E}_Z}{dX} = \frac{d\beta_Y}{dt} \quad \text{and} \quad \frac{d\mathcal{E}_Y}{dX} = \frac{d\beta_Z}{dt} \qquad [VII - 6]$$

Maxwell's fourth equation is based on Ampere's principle, which states that the total magnetic flux around a closed curve is equal to the permissivity (μo) times the current (I) linking the closed curve:

$$\beta \, dl = \mu_0 \, I$$

By the application of a rather complex logic, Maxwell arrived at the differential form of this principle:

$$\frac{d\beta_x}{dY} - \frac{d\beta_Y}{dZ} = \frac{\mathcal{E}_o \mu_o \, d\mathcal{E}_X}{dt}$$

$$\frac{d\beta_x}{dZ} - \frac{d\beta_Z}{dX} = \frac{\mathcal{E}_o \mu_o \, d\mathcal{E}_Y}{dt}$$

67

$$\frac{d\beta_Y}{dX} - \frac{d\beta_X}{dY} = \varepsilon_o \mu_o \frac{d\varepsilon_Z}{dt}$$

By dropping the terms containing **d/dY** and **d/dZ** he arrived at his fourth equation:

[VII-7]

$$\frac{d\varepsilon_z}{dt} = \frac{1}{\varepsilon_o \mu_o} \frac{d\beta_Y}{dX} \text{ and } \frac{d\varepsilon_Y}{dt} = \frac{-1}{\varepsilon_o \mu_o} \frac{d\beta_Z}{dX}$$

Based on these four equations, Maxwell concluded that:

The first and second equations state that $d\varepsilon_X$

$= d\beta_X = 0$ therefore ε_X and β_X must be constant in both space and in time so that only the transverse components ε_Y, ε_Z, β_Y, and β_Z can contribute to the any wave motion. Because of

68

this, Maxwell concluded that electromagnetic waves must be transverse waves.

If the wave is plane polarized with $\mathcal{E}$ in the **Y** direction, then β must be in the **Z** direction. If it is plane polarized with $\mathcal{E}$ in the **Z** direction, then β must be in the **Y** direction.

Differentiating the second part of [V11-6] with respect to **X** and the first part of [V11-7] with respect to **t**:

$$\frac{d^2 \mathcal{E}_Y}{dX^2} = -\frac{d^2 \beta_Z}{dX\, dt} \quad \text{and} \quad \frac{d^2 \mathcal{E}_Y}{dt^2} = \frac{-\,d^2 \beta_Z}{\mathcal{E}_o \mu_o\, dX\, dt}$$

thus:

$$\frac{d^2 \mathcal{E}_Y}{dX^2} = \mathcal{E}_o \mu_o \frac{d^2 \mathcal{E}_Y}{dt^2} \qquad \text{[VII-8]}$$

which Maxwell recognized as the equation for

a transverse wave traveling along

the X-axis:

$$\frac{d^2y}{dX^2} = \frac{1}{v^2}\frac{d^2y}{dt^2} \qquad [\text{VII-9}]$$

Comparing [VII-8] to [VII-7], Maxwell concluded

that the wave speed **(v)** was equal to $1/(\varepsilon_o$

$\mu_o)$ or 300,000 Km/sec.

By differentiating the first part of [V11- 6] with

respect to **t** and the second part of [V11-7] with

respect to **X,** another set of equations related to

β_Y, shows that the speed of the magnetic field is

the same as the speed of the electric field.

Since 300,000 km/sec was already known to be the speed of light in free space, Maxwell concluded that electric fields, magnetic fields, and light are all electromagnetic waves and, because of the polarization phenomenon, he also concluded that all electromagnetic waves are transverse - not longitudinal. Based on his own perception of his work, Maxwell's Theory is not specific as to the frame of reference to which the predicted speed is related. Instead it raises a question as to whether the speed refers to some special frame of reference which the observer can say is at rest so that an observer in motion (with respect to the special frame) would measure a light speed other than the constant 300,000 krn/sec. This is the question that, some maintain, prompted Einstein to begin his investigation that eventually culminated in his Theory of Special Relativity.

Maxwell's rationale for his conclusion (that electromagnetic waves are transverse waves) is difficult to accept since it is based on his first and second equations that describe the fields for static conditions in which the field intensities do not change. It is also difficult to understand why Maxwell assumed from the beginning, that he was working with plane waves. These two assumptions alone are enough to insure that he would be led to believe that he was working with a transverse wave to begin with, and yet the transverse wave is given as a conclusion. In addition, Maxwell's imaginary fluid is frictionless and, as such, cannot sustain the shear forces required for a transverse wave to exist. Also, since the fluid has no mass there cannot be energy transfer by wave motion.

Another question arises when we consider Maxwell's assumption that the electric and magnetic fields are perpendicular to each other and to the direction of propagation of the electromagnetic wave, and that this condition is responsible for the phenomenon of polarization in the visible light portion of electromagnetic spectrum. This assumption and the resulting conclusion cannot be justified in a study that is supposed to be based strictly on empirical relations and abstract analogies.

Still another problem appears when Maxwell, without justification, equates his resulting equation for the acceleration of the electromagnetic wave with the equation for the acceleration of a transverse wave in a flexible string. Recognizing, however, that Maxwell was not building a theory but was, instead, using analogy, as a means to construct a mathematical tool to predict and to solve electromagnetic problems, we may accept these questions as unimportant so long as we, also, do not try to build an electromagnetic theory around his equations

Unfortunately, Maxwell's electromagnetic field equations have evolved into a theory of electromagnetism and, as such, must be reviewed. One obvious, but apparently ignored, implication of Maxwell's equations is that all electromagnetic waves (including light waves) may be generated in such a configuration as to make possible the propagation of wave speeds in excess of 300,000 km/sec.

This becomes apparent if we consider the equation that describes any sinusoidal wave, in terms of **t** and **x,** having an amplitude **A,** and traveling in the **+X** direction such that:

$$Y = A \sin 2\pi f(t - x/v). \qquad \text{[VII -10]}$$

In the special case (referred to by Maxwell) of the sinusoidal, transverse wave where the time **(t)** equals zero when the vibrating source is moving upward through the origin, the actual form of the sinusoidal transverse wave is as readily apparent as if in a snapshot.

In order to derive the expression for the speed of any sinusoidal wave, we begin with the equation for a sinusoidal wave traveling in tlie positive direction, as in Equation [VII-10] and find the first derivative of **Y** with respect to time **(t)** when **x** is constant to be:

$$dy/dt \quad = \quad 2\pi fA \ \cos 2xf(t - x/v) \qquad \text{[VII-11]}$$

Then find the first derivative of **Y** with respect to **X** when the time **(t)** is constant

$$dy/dx \quad = - (2\pi f/v) \ A \cos 2xf \ (t - x/v). \qquad \text{[VII-12]}$$

Combining [VII-9] and [VII-12] results in:

$$dy/dt \quad = \quad v \ dy/dx$$

and setting v equal to the constant c (=300,000 km/sec):

$$dy/dt \ = c \ dy/dx$$

76

which tells us that **dy/dt = c** only when **dy/dx = 1.0**

and thus:

$$dx/dt \ = \ dy/dt \ = dz/dt \ = \ c \qquad\qquad [VII\text{-}13]$$

The significance of equation [VII - 13] is that the velocity of the wave, in all directions perpendicular to the x-axis, is the same as the velocity of the wave along the x-axis that indicates that:

a) the wave is a longitudinal (compression) wave,

b) the wave velocity in the **x** direction may be greater or less than **c**, (depending upon the value of **dy/dx}** ,

c) the wave velocity in the **y** direction may also be greater than or less than **c** , depending on the value of **dy/dx,**

The longitudinal wave is subject to the laws of polarization, (in any conducting medium, the ability of the material to support vibrations in two of the three mutually perpendicular axis). In the case of light, any conductor that will support vibration in two of the three mutually perpendicular axes will conduct the **y** vector component of **y-z** vibration only, along the **x** axis, creating a polarized wave traveling in the **x** direction. A similar conductor, placed on the path of the polarized ray, will either conduct, partially conduct, or block the polarized ray depending on its orientation with respect to the polarizing conductor.

In the case of electromagnetic waves, other than light, the polarizing device may well be a simple electrical primary conductor. When an electric current is passed through the primary conductor, the electrons could, conceivably, generate a shock wave ("electromagnetic"- wave) in the ether surrounding the conductor. As the shock wave travels through the ether it may be sensed only by another conductor oriented, in space, parallel to the original conductor. When this happens, the shock wave may set the electrons, in the second conductor, in motion as an induced current. It is obvious, therefore, that this alternative interpretation of Maxwell's equations is as valid as Maxwell's own interpretation.

Joseph L. Poggie & Robert A. Poggie, Ph.D.

CHAPTER VIII The Universe and the Big Bang

The concepts presented in this book are based on the premise that Newton's Laws of motion are the only true laws of physics. There are no exceptions or modifications and his laws of motion are accurate and universal.

With this in mind, the authors selected certain important empirical equations, such as the theories of gravitation and of electro magnetic propagation as well as experimental data and observations gathered by scientists and engineers over the years. This information was organized and examined as to the relationship of various theories to each other, to the gathered information, as well as to the principles of Newtonian Mechanics.

The results suggest that, some sort of violent explosion, or 'Big Bang', occurred and started the process of forming the universe, as we know it today. The information also permitted scientists to reconstruct a plausible history of the universe, from its beginning until now.

It is generally accepted as fact that, a few moments after the big bang, the universe consisted of a huge, isotropic, mass of very small bits of primary matter immersed in a cloud of even smaller bits, (gaseous matter or ether), all of which was expanding uniformly in all directions. As the entire mass expanded, in accordance with the laws of expansion of an isotropic mass, the holes in the ether, (in which the small bits of primary matter were imbedded) also expanded creating a vacuum like void around each tiny bit of primary matter. As the vacuum voids increased in size, ether from the periphery of the voids rushed in toward the center of pressure each void. Any disturbance or irregularity within the entire expanding isotropic mass would cause two or more voids to combine.

When the voids combined, their centers of pressure also combined and relocated in a new position somewhere between the two originals. The result would be that the inward flow of ether would now shift toward the new center of low pressure, which in turn would create a new force on each of the particles toward the new center of low pressure. In the case of two particles, they would be forced together to form one new, more massive particle. Where more particles are involved, they might combine in a grazing or off - set pattern (atoms), rotating about a newer center of low pressure. As the universe expanded further, the voids containing atoms combined in a similar manner to form molecules, then further expansion combined the molecules into matter, and ultimately into planets and galaxies within the universe, as we know it now.

The authors predict a logical continuation of this scenario, based on Newtonian mechanics. The universe would be expected to continue expanding, at an increasing rate, due to the force of the essentially constant pressure of either inside the universe vs. the absolute zero pressure outside the universe. With further expansion, the universe will become one void containing all of the primary matter, plus most of the ether in universe. At this time, the void and the universe would each continue to expand, while the galaxies would now be forced closer together until all of the ether outside the void began rushing inward, forcing the galaxies together. The galactic motion would be a rotating pattern; with increasing inward ether force on all the matter in the universe being balanced by the pressure of the ether within the center of the void plus the centrifugal force of all the rotating matter. When balanced, at this point in time, the contraction of the universe would stop and the universe would then begin to expand all over again.

The Unified Theory of Mass, Gravity, Light & The Big Bang

Of course, all this is speculation. The real truth

- only God knows.

Joseph L. Poggie & Robert A. Poggie, Ph.D.

REFERENCES

1. Ashby and Miller, <u>Principles of Modern Physics</u>. San Francisco: Holden-Day Inc.,1970.

2. Born, Max., <u>Einstein's Theory of Relativity</u>. New York: Dover Publications, 1962.

3. Niven, W.D, ed, <u>The Scientific Papers Of James Clerk Maxwell.</u> New York: Dover Publications, 1965.

4. Poggie, Joseph L., <u>Special Relativity Tested</u> Bloomington IN: 1st Books Library

5. Shortley and Williams, <u>Elements of Physics</u>. New York: Prentice$^{©}$ Hall Inc.,1971.

6. Simpson, Thomas K., <u>Maxwell on the Electromagnetic Field</u>. Rutgers University Press, 1997.

Other articles and publications on Gravity, Light, Relativity, and the Origin of the Universe,

About the Authors

Joseph L. Poggie majored in Applied Physics at Hofstra University, on Long Island, N.Y., and at The Florida Institute of Technology, in Melbourne, Florida. He has more than forty years of Engineering Research and Development experience most of which is in the aerospace industry and in the U.S A.A.C. He has spent the majority of the past fifteen years of his retirement developing the theories presented in this book, and those presented in "Special Relativity Tested", published in 2000. He also has had many articles published in technical publications and magazines.

Robert A. Poggie majored in Mechanical Engineering at Vanderbilt University in 1984 and received Masters and Doctorate Degrees in

Materials Science and Engineering from Vanderbilt University in 1986 and 1992. He has over twelve years of experience in orthopaedic research and development, and is currently employed as Director of Applied Research by Implex Corp., located in northern New Jersey. He has over fifteen published articles and sixty abstracts from his work in orthopaedic research and development.

Printed in the United States
47574LVS00001B/107

9 781403 370877